Table of Contents

Chapter 0: Preface

Damn, another book on lean tools? Maybe but more of an approach or methodology rather than just tools. This book is sort of a sequel to the book "Facility and Process Design Using L3P" released in 2018. That book covered the mostly fictitious story of Dynet, a global manufacturing company with a plant in the Midwest USA, Breckenridge, that was struggling to meet growth and productivity among other metrics and faced the dilemma of having an opportunity to in source production other internal plants as well as suppliers but was constrained on space. Breckenridge was operating in more traditional, batch style flow environment and space was not being used strategically. This allowed the concepts of Layout 3P to be used successfully to free up space and improve macro level flow in the end.

Towards the end of that story, it began coming clear that to execute some of the L3P plan actions, some manufacturing cells would have to be redesigned and the footprints reduced. The team also noted the opportunities to improve quality as well as throughput time and productivity in these cells as well. This is where the last chapter, Design for Flow, introduced the concept of DFF, and as the team began planning to utilize that approach in their assembly area.

That's sort of where this book will pick up. Continuing that story line but now focused on more of the micro aspects in the factory using an approach similar to the L3P method but scaled down to the micro level. As with L3P, we view this as more evolution rather than revolution. None of the tools involved with DFF area magical and many have been around for decades - with standard work, PFMEAs, process flow charts, being a few examples. What is different is the method to bring it together and approach the task at hand to in the end get the best design to enhance safety, quality, throughput, and productivity while engaging the entire team.

Again, this may seem like a bit shorter book but it's focused. The intent is to be more of a field book, in a story format followed by perspective and application wrapped up with an action summary in each chapter. We aimed for a practical overview with minimal fluff to get the outline and concept to you, the reader, in a story like but impactful manner while also providing access to resources for additional information as well as supporting tools virtually if desired or needed.

I personally along with a few teammates have applied and refined these concepts for many years in manufacturing settings and have trained many people on the concepts with a broad amount of successful outcomes. Again, the tools aren't new, just the approach maybe. Personally, I have a passion for manufacturing and have grown up in it and will always be an advocate for manufacturing in our country. I believe integrating technology and lean thinking with an emphasis on people will be key to our future competitiveness in America, and applied tools like this in my view are designed to help us improve, in this case with respect to space and material flow. I hope you feel the same. Be sure to check out my personal blog shown below and follow for access to additional content as well as supporting tools and materials related to L3P, DFF, and other topics. Enjoy, keep learning, and always compete.

Brian D. Summerfield

February 2019

Published in the United States of America

First Edition, 2019

ISBN: 9781794401082

Published by TwoEighty3 IC LLC

202 Main St, #190

Pine Bluffs, WY 82082

www.briansummerfield.com

Chapter 1: Introduction

The book continues a mostly fictitious story line about a fictitious corporation, Dynet, and one of their sites in the Midwest US, Breckenridge. The previous book on Layout 3P walked the reader through the pressing dilemma to free up some space to in source work and potentially fight to keep their plant not only open but thriving. Succeeding towards the end of that story line at least with their proposed plan, this story continues to tell the story using the Dynet Breckenridge backdrop, but now the team is faced with actually executing some of the L3P action plan in the facility. To do so, they must reduce the footprints of various departments and work cells while also imploring flow and hopefully (yes, hope is not a good strategy), performance in these areas.

The chapters begin with the story narrative, with each chapter then providing a perspective and application section to provide more insight into the concepts presented in the narrative. The chapter then ends with a brief action summary to provide a detailed list of steps or concepts laid out in that chapter.

The setting and characters, all fictitious by name but not by personality for some individuals, include the following for reference, many continued from the L3P book but with some new additions in this story.

Dynet – A 110 year old corporation that manufactures HVAC equipment, with three divisions including Mechanical, Thermal, and Controls.

Breckenridge – Midwest US site in the Dynet Mechanical division. Site was an acquisition eight years ago by Dynet. Lean implementation has been slow, culture is somewhat poor, and performance overall has been mixed. It is the second largest site in the Mechanical division by revenue and size.

Greg – Dynet Breckenridge Plant Manager. He has been at the site for over 10 years, first in operations management, then the plant leader role for the last 3 years. Greg is long time operations management veteran, having worked for a number of companies, and has a more traditional view of manufacturing. While he does support "lean", he does so with more reluctance and skepticism and feels compelled to at times.

Ben – Breckenridge Facilities Manager. He has been with the site for over 18 years, all in facilities and maintenance roles. Very opposed and outspoken to newer ways of thinking and doesn't view "lean" as useful, often referring to past attempts and failures at site.

Jeff – Breckenridge Manufacturing Engineering Manager. Started at another Dynet facility as an ME out of college and moved to Breckenridge two years ago as the ME Manager.

Progressive thinker who is strong advocate for lean and new technologies.

Chris – Dynet Mechanical Divisional VP of Operations. He is Greg's direct leader and driver of the dilemma facing the site, wanting a solution for the business. Believes in lean and tools, advocate, but driven for a feasible solution first and foremost.

Harold – Breckenridge's Materials Manager. Team member, who is somewhat reserved and sort of on the fence. He has been at site for over 10 years in materials roles, and is heavily influenced by Ben and often defers to Greg.

Maria – Site Quality Manager. Team member, who is a newer member to team in the last year, and a very progressive thinker. Can be quiet, as more introverted, but with right mix of people, provides good input.

Ace – Manufacturing Engineer. Less than 18 months out of college, and very progressive in thinking. Part of Jeff's ME team and very supportive.

Ellen – Site EHS Manager. Also newer team member, been with site for about one year. Heavily influenced by Greg, her manager, but also provides high energy and good input.

Brian – Mechanical Divisional CI Support.

John – Mechanical Divisional CI Support.

Joe - Dynet Breckenridge Assembly Manager. "Promoted" into the role from supervisor after his former boss was let go. The previous predecessor quit. Joe found out why soon after taking the role. Assembly is a mess.

Dave - Engineering Manager, Dynet Breckenridge. Been with company for nearly 10 years, knows many products including the legacy ones. More focused on product development - almost to point of competing with manufacturing. Will support initiatives if upper leadership drives it or if convinced right thing to do.

Vivek - Design Engineer, Dynet Breckenridge. Younger, very smart engineer who has been at Breckenridge for last two years after a rotational program took him to three sites in three years. Has picked up products very well and eager to improve them and has a special respect for manufacturing as his father owns a shop in India.

Hazel - Assembly Operator, Dynet Breckenridge. Been with company at this location for 20 years. Knows various products well, highly enthusiastic but strongly opinionated in doing things the traditional ways, even before Dynet acquired the site. Part of a long list of workers ("baby boomers") who have been with the business for a long time and approaching retirement in the next couple of years.

Brad - Assembly Operator, Dynet Breckenridge. Been with the site for 3 years, younger generation of workers (Gen Y) on the team. Eager to learn, eager to change things. High energy but can be outspoken by the longer-term workers at the site.

Chapter 2: Set the Stage - The Situation

Joe pulled into the parking lot of Dynet Breckenridge just after 5:30 in the morning. He sat in his truck for a few minutes just staring out into the marsh behind the factory parking lot. His shift technically starts at 7:00, but as the area manager for assembly, he has learned to show up early for the last year since he was "promoted" into the role. He technically wasn't promoted. No pay raise, no fanfare, no move off the shop floor where he sat. The last two assembly area managers left – one by choice and the other, his most recent boss, was let go. And Joe found out why soon after Greg informing him of the change.

It was a hard role to fill for sure. 10-14 hour days plus Saturday was a common theme. The daily and weekly list of problems never seemed to subside. Part shortages from suppliers, late parts from machining, quality issues, equipment problems, and inability to deliver product on time, compounded by low productivity, low morale, and the seemingly constant need for temps and OT.

Inside, Joe made his way through the shop stopping to check on some of his assembly cells. In less than 15 minutes, he was stopped 3 times by a team lead and two different operators with "urgent" issues. He looked at the cells in front of him as an operator complained about something. He noticed large batches of parts between operators. Many of the cells had several operators - and yet still struggled to meet output. The common theme had been to just add operators. It clearly wasn't working. There had to be a better way.

"Joe, how are the cell moves coming along?", asked Greg, his plant manager who had walked up. Joe has been tasked with moving, and trying to consolidate, half of his 8 assembly cells in the next two months to complete the Layout 3P plan the site team had completed last month. "Coming along Greg. We moved cell 1 last week but that one was easy. We were supposed to move cells 2 and 3 starting this week but have to shrink the size by at least 20%. We just aren't seeing how right now," responded Joe flatly.

"The assembly moves need to stay on path so we can move some of the milling machines on schedule Joe," replied Greg flatly and then added, "The CI guys are helping us prepare for an event that supposedly will help us optimize our cells a bit and get the reductions needed. They talked about it during the layout event and we have had a couple prep calls with them last few weeks. You have been working with John and Brian to get them the info needed, right?" Joe responded, "Yes. We have been working to layout the products, volumes, and the routings, which has not been easy. Hopefully they are getting something understandable."

Joe wasn't a skeptic towards "lean", he had seen some positive results from it when working at a competitor in the past. But here at Dynet Breckenridge, it was different. Lean here wasn't encouraged like it was at his last company. And he was overwhelmed as were his people. For them, it was about survival day to day. It was about getting the gauge back to zero each day. But he knew deep down, that this could potentially help them, especially after seeing the energy around the plans the team came up with in the layout 3P event.

- - -

We were planning to kick off the event early that Tuesday morning. John and I could feel the positive energy through the team from the L3P work in Breckenridge still from our previous visit last month to help prepare for the event we were now kicking off. We had arrived later Monday and came in to tour the facility with Greg and Ben. The team had been doing some great work. Machines were moving, some infrastructure work was being completed, and things were progressing to overall plan.

But Ben and Greg were a little more frustrated when we got to the assembly area. Ben lamented, "It's easier to move equipment from here to there or to have electricians come in to do some power work. But this," he said looking across the complex pattern of benches, inventory racks, air lines, and other items seemingly strewn about in assembly, "is where Greg and I are struggling a bit with the team to figure out how to get this organized, reduced by 20% per the plan and moved into the back northwest corner per the layout."

"It is tough to see what the hell is going on here for sure," John said. We had seen this area last month, walking through it and helping the team build a PQPR or Part Quality Part Routing matrix that would help get their products or value streams better aligned. It was a painful exercise to say the least but has helped them begin to see things in a more organized fashion for the future. It also helped us define the core products that will go into this first cell we are working on now.

"The plan now that we have the PQPR and did some mapping with you guys, is to do this first event with you on the XS product cell. This is your highest volume, and highest margin, product line and fits nicely into one cell.' I said. "You guys can then take this knowledge and hopefully with support begin doing the other cells. The PQPR indicated the need for 5 cells in total," John added.

"There are roughly 30 people total in assembly today, and the XS product, assembled in three different areas today, requires about 10 of those people," added Greg and continuing, "we obviously need to get the footprint reduced overall for XS but any improvements we get are much needed. It's our highest past due family every month, we have had numerous quality issues and it's impacting our sales." "That's where this effort can help. Obviously getting the footprint reduction but also improvements in safety, quality and throughput as we have been saying," John input.

We continued to walk through the factory to see some progress as well as a general gemba walk to see some opportunities and discuss some key points. We ran into Joe, who was still there into the second shift. "Hey Joe, you remember John and Brian from headquarters?" asked Greg. "Sure do. We worked on the PQPR and mapping last month. Hope you guys got everything you needed. I am looking forward to the event," Joe said. John and I smiled and nodded, but I could tell his enthusiasm was a bit manufactured. The truth was Joe mine as well have been in another state when we were here last month. He was pulled in several different directions, his focus was limited, and you could tell his energy and morale of him and his team was low. It wasn't that he didn't want to do this, he is just overwhelmed. In addition to the technical challenges the plant was facing, you could feel undertones of cultural and people issues as well.

We exchanged some banter and then kept moving, saying we would see him in the morning and continued to complete our walk through. Later, John and I were chatting at hotel as we prepared. "It seems like we have some engagement from Greg and Ben now," John said. "Yes, they have come around. I am concerned about Joe and his team. They are swamped and just need air so getting them engaged will be key," I added. Little did we know other challenges would surface that would rival the struggles with we had with Greg and Ben during the early L3P work.

Perspective and Application

Design for Flow or DFF for short, can help you improve your processes while optimizing your footprint and throughput as well as drive breakthrough improvements in safety and quality. It uses a collection of different tools, nothing radically new, but in a similar way as Layout 3P, brings them together in a unique way to approach the design process. In this situation at Breckenridge, the team needs a method like this in order to attack their assembly area, reducing the footprint as needed but also to improve the operation which was under performing.

Preparation is important for DFF. Isolating to a specific product line or work cell is important for focus. The PQPR (Part Quantity, Part Routing) matrix and value stream mapping can help. If a cell exists already, it can be re-worked using the DFF approach. It can often lead to dramatic improvements as old methods and standards are not only challenged but rewritten. On the flip side, DFF can also be an excellent approach to new product introductions that require new processes and innovative thinking. Ensuring that a well-rounded team is planned cross functionally, having components available ahead of event, and having some pre-work data collected are important as well.

Chapter 2 Action Summary

1. Select the focus area for the effort. A key product line or work area (cell) is suggested. Ideally an impactful one that requires some improvement, or a new product requiring a new process.

2. A PQPR (Part Quantity Part Routing) matrix (reference example in appendix) can help better clarify product groups, work cells, and overall flow paths. This may be helpful in choosing focus areas for DFF efforts.

3. Doing value stream mapping to help identify the current state macro level flow can be helpful as well to identify some opportunities. A more detailed study of the flows will be done in the DFF event.

4. Think about the team composition - strive for a well-mixed cross-functional team including design and product management, which are important especially early in the event.

The DFF prep guide can provide some additional detail for preparation and lead up to a typical DFF event as well as the post activity. This can be referenced in the appendix and is available as a supporting add on.

Chapter 3: The Product Study – F^3 Review

The room started filling up around 7:30 AM for the event, very different from the L3P event a while back, where people were still filing in past 8 AM. John and I had prepared the room early as well as last night and we would begin with some training on what we were about to embark on over the next few days or weeks for that matter. "Good morning", I said to various people as they came in and grabbed coffee or bagels. There was a different energy in Breckenridge today. John and I had talked about it last night. It was more positive than our previous visits but still a cautious apprehension. The site was not out of the woods - they had an incredible amount of work to do to execute the L3P plan and this work here was key. The assembly area needed to be reduced roughly 20% overall from the 15,000 square feet it occupied today.

John got things kicked off, "Good morning. Hope everyone is doing well and welcome. The first thing we are going to do is introductions. Give the group your name, your role, and how long you have been with Dynet." We cycled through everyone and then John and I made our introductions. "We are going to do a little exercise but after we review the event plan. We know we are here to do some Design for Flow work in the assembly area. We will be focusing on the XS product group, your highest volume product and one that is critical to the site. Our objective is to create one, maybe two if needed, work cells to produce these assemblies in the safest, highest quality, and most productive manner," I said.

We continued to step through the charter. "We want to find a minimum of 3 safety improvements, minimum of 3 quality improvements, and improve productivity by 50% minimum. Additionally, we want to reduce the current assembly footprint for XS by 25% minimum," John stated confidently. "Does anyone doubt we can achieve this?" he added, more as a statement but in question form. The room was quiet until Greg spoke up. "I damn sure think we can. The XS assembly is a mess. It is done in 3 different areas, our quality is poor, and we are always behind. The only way to go is up."

I added, "Another important aspect of this is to teach you guys this approach so you can begin using in other areas of assembly and the factory. You will need this method to help reduce your footprints but also to improve your operations." We finished cycling through the charter and answered a couple of questions. John then said, "everyone take a Post-It. Think about what is limiting you today in achieving what we want to do in improving assembly and reducing the footprint as we need to do." After a few minutes everyone finished writing. "Now, fold the paper a couple of times," he added. "And now tear the paper up two or three times. There is nothing limiting you. We give you permission to be creative and develop solutions to make this happen. Do not limit yourselves in this event." With that, everyone tore up the papers and we were off. John and I tag teamed the initial training overview slides, which we kept relatively short to focus on application learning. We had learned to use micro-learning approaches in the slide presentations - present the concepts just ahead of discussion or application in small chunks. We answered some questions and then took a quick break.

- - -

"We start with the product or the F^3review," John declared. "We want to understand the product as far as form, fit, or function because it will help us design a better process," he added. John and I had noted, mentioning something to Greg, that no one from engineering was at the kick off. And they still weren't in the room. Neither David, the Engineering Manager, nor Vivek, the Design Engineer were in the room. Greg had not come back from break and was obviously trying to find them.

John continued talking about the importance of the product. We had got a feeling that Dave was going to be a handful as he had been on a few of the prep calls, reluctantly. He was negative towards what we were doing, saying the products should be off the table as they don't want to change designs and his team shouldn't be involved, as they had to focus on "new products".

Vivek was a breath of fresh air when we met with him against Dave's wishes really when we were on site last month. He was super knowledgeable about the products and very helpful... Until Dave came around. Then Vivek became quiet and less willing to cooperate.

Greg, Dave, and Vivek finally emerged. "Good morning guys," said John trying to stay optimistic. Dave and Vivek both waved their hands and grabbed some coffee, taking seats in the back. "Dave and Vivek are going to do a product overview for us on the XS so we can learn and better apply in our process design later," I said. Dave followed my last word, "Vivek is going to give an overview. I can fill in where needed."

Dave had been with the company for nearly 10 years. He knew the products well and was certainly committed to developing new ones, maybe a bit too much. He reported outside the site to a VP of Engineering at the Mechanical division level. Him and Greg were essentially peers on site. He had a team of 12 engineers in the site, many focused on NPD but a few towards sustaining existing products.

"Good morning. I am Vivek, design engineer, and been with Breckenridge for the last two years. Before here, I was with Dynet Thermal for a year in Waterloo and did a couple of other rotations before that," Vivek stated. Vivek then began handling and speaking to the various components of the XS laid out on the table. He began with the overall assembly and the function, what was important, and then down to the component level on what was critical to form and fit. He knew the product well and would be a great asset in this; when he was available.

The XS was essentially a mechanical ball valve that could be manually actuated or automated. It was used to control air flow in HVAC systems. It was an older Dynet design and therefore did not have a Design Failure Mode and Effect (DFMEA) analysis. It did, however, have a Process Failure Modes and Effects Analysis (PFMEA), the team had put together, although quick and crude. It came in four different sizes with some different configurations in seal types, material, and add on configuration option such as the actuation. The seals were obviously critical and could not be damaged, which was a major quality issue in the past. The torque of the stem securing the ball was another. Varying this affected actuation torque and valve performance. An example of the valve is shown below in figure 1.

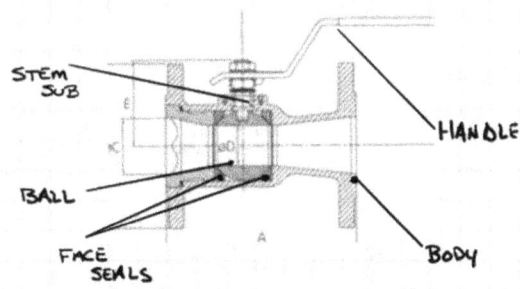

Figure 1: Overview of XS Ball Valve Type

As Vivek walked the team through the valve function, Maria the quality manager chimed in adding some comments on critical points from the PFMEA. "Areas to watch carefully that have burned us recently as well are the ball seals being damaged which is a leak path, the stem seal being over torqued, or not enough torque being applied which affects the packing of the valve and the actuation torque," Maria commented. This was positive as quality inputs are key at this stage.

As they continued, Dave got up and left the room with his notebook and phone. Before he left the door I noticed Dave give Vivek a "hurry up and finish sign" that was noticeable. Vivek wrapped up quickly after that and noted he would have to step out for some meetings. It was a good break point for the team after a couple of questions were answered.

During the break I asked Greg, "When will Vivek and Dave be back?" "Not sure. Dave gave me a real hard time about being in this event and I actually had to call Chris who called Dave's boss. It wasn't easy," answered Greg flatly. "We really need engineering participation in this. Vivek did a great job and it would be really good if we can get him in the event with us," John added. "I will see what I can do," said Greg as he pulled up Dave's number on his phone.

After the break, we had the team continue reviewing the PFMEA and the parts to see if anything to update or revise. In this case, the team had a working PFMEA to use as a baseline and adjust. Sometimes, maybe more often, there is no DFMEA or PFMEA, and the team may have to create at minimum a PFMEA, but not in this case. The team had a good understanding of the design, and the two operators we had in the event, Hazel and Brad, both commented they had been assembling these for years and never knew this much about the product. This was exactly what we wanted to hear at this point in the process. It was likewise great learning for the manufacturing and quality engineering team members.

Perspective and Application

The product overview is the preferred start to a solid DFF effort. It is essential that the team is grounded in understanding the form, fit, and function of the product. Amazingly, in many cases we often find that operators, manufacturing engineers, and front-line leaders do not understand this critical information and are expected to build and test high quality products as efficiently as possible. It is always eye opening to have the broader team learn and understand the product better and know what is most critical to it for customers.

This information will become very useful as the team begins to move through the next stages of the DFF process. In some cases, it can lead to completely different ways of thinking on how to build the product and this innovation is exactly what needs to be capitalized on. The best people to provide this overview area typically design engineers and/or product management experts who know the product and the end use applications very well. Ideally both can be present and provide the overview together as well as answer questions from the team.

Chapter 3 Action Summary

1. Assemble a cross functional team comprised of operators, front line leaders, manufacturing engineering, quality, design engineering, materials, safety, and if possible, product management.

2. Have parts and assemblies available for the event. Additionally, having customer information and end use applications is beneficial as well.

3. Have design engineering/product management provide a deep overview (F^3) of the product form, fit, and function for the team as well as answer questions.

4. Cross reference this overview with the design failure modes and effects analysis (DFMEA) if available. **If one is not, highlight the critical to quality characteristics the team should take note of and why.** In essence, doing a truncated design review to identify critical characteristics.

5. Drive through to a process failure modes and effects analysis (PFMEA) for the process. This may involve reviewing an existing

one for updates and learning or having to create one. In any event, developing a good PFMEA and talking through with the team is key to understanding the product and process before moving ahead.

Chapter 4: The Process Study

After the break, John rounded the team up back to their seats and began to give a brief training overview on what was to come. Vivek had come back with Greg which was encouraging to see. Dave was still not around. "Now that we have focused on learning more about the product, we need to transition now to the process and begin to learn how we do it today," John stated. "As we did for those who were with us in the Layout 3P, taking the time to understand the current state of the process is good for grounding us and helping see with more clarity where we have gaps and what is possible in terms of opportunities. It is another way, just like with the product overview, to trigger our imagination and ingenuity. But now it is time to pivot." I added.

"We are going to cover some basic lean concepts to help prepare us for what is to come," John said as he began presenting. We covered a basic deck showing the purpose of lean to reduce waste, streamline lead times, and ultimately maximize our value add to the customer. Particular emphasis was placed on the breakdown of work in elements. "We have three types of work we need to be keen to, value add which are customers love, waste which our customers don't love, and non-value added activity that we either have to do or can't entirely eliminate today," I interjected as John finished the slide. "What are some examples?" Brad, one of the operators asked. To which John replied, "Great question Brad. We will do a quick exercise but for now, a value-added task would be if I am joining two parts together via welding, the weld task would be an example of value added. If the two parts I had to weld were far from me and I had to walk to get them, the walking would be waste. If I then had to enter some data into our ERP system after completing the part, that is an example of non-value add but required."

We continued the training also placing more emphasis on the various forms of waste and how to spot them. We placed great emphasis on some of the forms more often encountered at the work cell or micro level including motion, but also transportation. "A good rule of thumb for motion is to focus on people movement, such as walking or even movement of hands. When thinking about transportation, focus more on the movement of materials, in our case flow of components to the cell and within the cell," John added also stating that this would be important as we did the process study. Vivek had some great insight, sharing his perspective on wastes he sees on the production floor when he goes out there, but was interrupted when Dave appeared in the doorway waving at him. With that, we lost Vivek again, with Dave being part of the problem.

We then did a quick exercise that does a good job of outlining the work elements as well as various forms of waste in a simple process. After some clarifying question and answer after, we were ready for paper airplanes we told the group. "Paper airplanes?" Hazel asked. "Yep, after the break, we are going to make some paper airplanes to learn and have some competitive fun," I said. I heard Mark, the assembly supervisor who had been with the company for 25 years mutter something to Hazel to the effect of "This is such as a waste of time. Something new - making paper airplanes." He then rolled his eyes and walked out for the break. It was becoming more and more clear who the challenges would be this week and beyond.

"We are going to build some paper airplanes to give you guys some perspective on what we have been talking about and focus on two things. One being the aspect of batch production, and two being the ability to observe a simple process and do a process study," John told the team. We broke the group of twelve into two teams of six. "We need two builders, one person to map flow and tools, one person to note wastes and opportunities, and the other two to do work observations and times," I said as we prepared the stations.

The two builders were spread out a bit, and the "raw material" was placed a good distance from operator 1. The instructions (or "orders") were to build batches of 6 planes, so operator one would have to complete his steps for a batch of 6 before moving along. Operator 2 would then have to finish their steps and complete a batch of 6 and move to inspection for further processing, which again was a bit of a walk from their station.

"While they are doing the building, the other four team members are observing and documenting what you see," John added. "Break down the work elements, then record times. Map the material and people flows, and record any observed wastes or opportunities," I input.

"Shouldn't we be out on the floor doing this?" Asked Mark, the supervisor. "Yes, and we will soon enough. Since these are new concepts, we want the team to learn in a controlled environment on a simple process. We have found this to help calibrate people and then prepare them to do the work on the floor," John answered. Mark sort of blew off the response. We could tell he was someone who if he had the choice would not be in this. As a front line leader, Mark would be a critical part of sustaining anything put in place, so his engagement would have to be monitored.

After conducting a few test trials, we let the teams know we needed 36 planes in 6 minutes. Some gasps came from a few as the way the process was setup it would be hard to do. It was a competition too and first team to either get 36 or have the most **quality** planes wins. The round was intense as they usually are but the teams both struggled. The front operator is overloaded with work, meaning the second operator has a lot of waiting time due to the imbalance - another demonstration point.

John and I moved between the teams focusing on the four for each who was doing the observations and mapping. We made sure they understood what to look for and recording proper elements, times, and mapping the right flow paths. Jeff, the ME Manager, was the waste observer for team 1 and was doing a good job seeing potential opportunities to improve. He was also keen to point out how Ben, the facilities manager, could work a bit faster.

Ellen, the EHS manager, was the other team's waste observer who was struggling to see a bit so John stood with her and observed to point things out. The time observation teams struggled with the batching concept at first but figured it out. "You will very likely see batching out in the factory later. That is why figuring out how to observe it here will help," John said.

After the 6 minutes were up, we stopped the teams and worked to do a debrief. Team 1 had a single 6-piece batch complete that were acceptable, and team 2 had zero - with one batch being rejected due to quality. "What did we see?" asked John. "Each team has 10 minutes to debrief internally and summarize your findings - times, balance, opportunities, and such," I added. "Don't focus on discussing improvements - that will come later. Understand the current state and what opportunities there are," John added.

Team 1 went first, and Jeff spoke, "We saw a lot of motion waste, walking. But also, in the motion to actually do the work, it didn't look efficient, mainly due to the batching. Operator 1 had too much work and there was waiting by operator 2. No balance and the times were inconsistent." "Great job," I commended. "Greg, how about you guys," I continued. "Not much to add to that. We saw a lot of similar things. In our case, we also noted that there was no standard work so each build looked a little different. We think this is what drove the inconsistency that Jeff mentioned. The batching didn't help - seemed to drive waiting," he added.

"Outstanding. This may seem a little goofy building paper airplanes but this simple exercise helps us learn a lot and prepare us for doing the same thing out in the shop in a more complex environment. We have found it to help greatly to take the time to do this ahead and also have a little competitive fun," I added. "The final step we want you guys to do here is draw an operator balance chart for your process. We talked about the balance charts in the training, and we want to you to draw one based on your process on a flip chart," I said.

Each team drew the time axis and plotted the two operators on there. They plotted their work elements along the y-axis.

"Now, identify the waste, NVA, and value add tasks with red, blue, and green so we can visually see the process," John continued. The teams discussed the work breakdown and then colored out their charts. There wasn't much red, but some additional blue, and small blocks of green here and there. We discussed a little about the benefits of operator balance charts and decided it was time to go observe. We would come back to the charts and our airplane operations later.

We divided the teams up again to observe the XS valve process. This was done in three areas of assembly today: sub-assembly or prep, build, and then testing was done in a special area. The parts were batched through each operation so we would need three teams to do observations. On a flip chart on the floor, we broke down the assignments, so each team member had a specific assignment covering element/time observations, people and material flow, tools and equipment, and observers for waste and opportunities.

It was 2:00 PM - day shift was turning over to night shift but fortunately this entire area worked a full second shift, so we would be able to observe for at least a couple of hours. We tasked the teams and armed them with clip boards, pencils, work sheets, and they would use their phones for timing. We made sure no further questions and that everyone understood what we were after. "Break down your assigned process into work elements. Once you have the process elements established, time them for multiple cycles. If you are mapping, follow the part flow and spaghettis map it, similar to L3P. Then do the same for the people flow, which will be different at times. Note the travel distances and any opportunities to improve. If you are on waste watch - find the wastes you see, and you should see many examples. Note it and note any opportunities to improve. If you have tools and equipment, we want to document everything we are using to build or test this product today," I confirmed before sending the team off and noting we would go until 5PM before wrapping up the current state and debriefing. And with that they were off.

Around 4:30, team members began heading back and we followed to guide next steps. "We want to draw out our operator balance charts as they are today on the wall. Before that, please huddle with your teams and discuss the element breakdown - clarifying waste, VA, and NVA tasks as well as confirming the lowest repeatable to get actual task times," John added.

The teams collaborated to summarize their findings and develop the balance charts. In total, we had two operators working in prep, three operators working in assembly, and one additional operator for testing. A total of six operators per shift, and the balance chart revealed imbalances in the batch-oriented process. You could see the large amounts of red and blue in the bars, with small amounts of green sprinkled in. There was a tremendous opportunity to improve here but we were just getting started. Additionally, the teams noted on flip charts the wastes and opportunities consolidated along the value stream, the equipment and tools, as well as the material and people flow.

"It's amazing what you see when you go do a deep study like this," Jeff said. "I walk by this every day, many times a day, and never realized it from this perspective how much waste we actually do," Joe, the assembly manager noted. Greg concurred along with the rest of the team, "I think it's clear to say we have some opportunities. It's like the L3P thing all over again but at a micro level now. I can imagine applying this type of work across the entire assembly area and the potential that is out there," he continued with some optimism in his voice.

With that we decided to call the day and asked the teams to clean up their summaries as we would continue working with them as the week progressed. "Tomorrow, we will continue moving into future state. Now that we have a better understanding of the product and the current state, we can begin thinking about how we can improve it," John noted.

Perspective and Application

While the product overview is critical to understanding form, fit, and function, the process study portion is the core of the DFF. It provides the current state basis in terms of facts and data that is so key to truly understanding where the process is today and the opportunities to improve it. Like the Layout 3P current state, some could view this as overly analytical and maybe even a waste of time. But after going through it and literally "seeing" what is reality and what opportunities there are, few will doubt the approach again.

Understanding of some basic lean aspects is key - various forms of waste, breaking down work elements, and understanding flow mapping are a few key concepts. Learning how to observe a process deeply and really break it down is important. We have used the paper airplane exercise successfully to help demonstrate some of the concepts covered while also providing a safe, easy learning environment so the teams are prepared to move to the floor in a live environment. We have done it both ways - with and without, and feel including it has helped participants, especially those new to doing this, greatly. You can also use other training exercises such as Lego builds, but we have found the airplanes is simple, quick, and gets the job done.

Observing and breaking down the actual process as a team with assigned roles is really the goal of the process study. The team must observe the work, break down the elements as it is run today, time them, and then use that information to create an operator balance chart. In parallel, team members should be mapping the material and people flow for the process, while also noting any tools, equipment, or supplies utilized. Another team member or two focuses on observing what wastes are present and potential improvement opportunities the team can debrief on later. The final outcome is a broken down process that is visually displayed so the team can discuss and debrief on to truly understand and help set the path forward to the future state. Each step through the DFF builds on the previous one and is carried through the effort.

Chapter 4 Action Summary

1. Provide a basic training overview focused on the various forms of waste, work elements, time observations, operator balance charts, and standard work basics.

2. Conduct an exercise on breaking down work elements into waste, non-value added but needed and value added.

3. Run round 1 of paper airplanes (or similar) simulation to help the team practice.

4. Conduct the process study, collecting the following:

- Work element breakdown and time trials to get accurate task times
- Mapping of the material and people flow end to end for the process
- Identify the tools and equipment used in producing the product
- Note any wastes observed and potential opportunities
- Draw the operator balance charts to visually show the process for review
- Breakdown the balance chart elements into waste, NVA, and VA

5. Debrief and discuss as a team the results and align on opportunities.

Chapter 5: Future State, Paper Kaizen, And Cell Design

At dinner that evening following day one, John and I debriefed on the day, how things went, and what the plan was for tomorrow. "Good energy today. Thought most were engaged. Vivek was solid when he was present. Dave is totally checked out, in and out and is pulling Vivek away physically and mentally," John noted. I agreed and added, "Mark is a tough one too. He is just sort of there going through the motions. Joe also seems distracted. I think he is so overwhelmed in the current environment, it's hard for him to see beyond the trees." "Or the biggest fires," John joked. We agreed that we were on track for the most part - wrapping up current state on day one is a good goal and allows us to move to future state on day two.

"Now the fun begins," John quipped as folks began taking their seats the next morning. All were present - except for the usual suspects, Dave, Vivek, Mark, and now interestingly Ben. Ben had been quiet on day one, did what he was asked, but hardly spoke. At times he would fade away but return relatively quickly. I had not thought much about it, but recalled it when he was not present this morning. "Where's Vivek, Mark, and Ben?" I asked Greg. "Don't you care where Dave is?" Asked Ace as he laughed. "He wasn't around much yesterday, so expect more of the same today. If we have Vivek, we should be okay," I pointedly answered.

"I'm not sure," said Greg as he filled his coffee cup. "Let me go on round up patrol," he then added as he strolled out the door. While he was hunting, we worked with the team to make sure the current state we did yesterday was clean and answered a few questions. "What's the plan for today?" Asked Ellen. "Today is all about future state and design," John noted. "The fun stuff because we get to innovate and come up with new ideas."

Greg emerged a few minutes later with Ben and Mark. "Vivek will be here at 9:00 according to him and Dave. Mostly Dave's input," Greg said as he rolled his eyes. We agreed to get moving and moved into some basic training on future state work to come. "Going to shift gears now and work on our takt time for the product family," I said, moving to the front of the room. I continued, "To do so, we need two elements, the annual demand for the product group and our available time."

This was a relatively straightforward product group with only 5 part numbers. Most of the part numbers were similar in assembly, just some of the components of the valve were different, but the process was similar and cycle times would be close, within the typical rule of thumb of plus or minus 20% in order to use a single blended takt. "Annual volume on the product family in total was about 196,000 - rounded up and adjusted with a 3% growth rate, which was par the past few years. We then divided this by 250 days per year and ended up with a daily demand of 784 units.

John covered the takt calculation, "Take the available time and with breaks removed, 410 minutes, or 24,600 seconds. We will start with one shift to see where it lands, even though you guys are running across two shifts. We then divide this into the daily demand of 784 units and end up with roughly 31 seconds per unit. This is our design or takt time, and represents how often customers are buying this product."

John went over to the operator balance chart the team worked on yesterday and drew a line horizontal across at roughly 31 seconds on the Y-axis. "This is the takt time for the product and our line or cell needs to be balanced to this in order meet customer demand," he said. The line cut across a few bars and yet some others were well beneath it. (shown in figure 2 below). "Do you guys see what the challenge is here?" I said pointing to the chart.

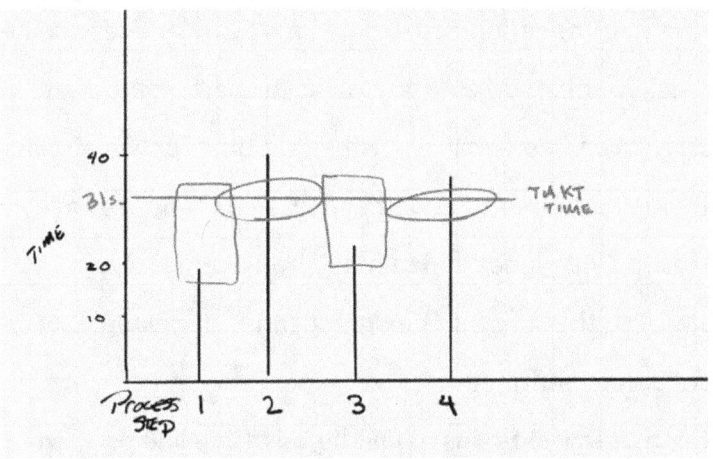

Figure 2: Balance Chart

Ace answered, "We don't have a balanced process. Some people are waiting and less work while others are producing too much and are over takt." "Absolutely correct," I said, continuing, "In this condition, we have no flow. It's our job to balance the line out and ensure we have continuous flow balanced to takt to meet customer demand. We are going to use a method known as paper kaizen to evaluate the process and see where we have opportunities to take waste out or minimize non-value added work. Our goal is to maximize the value-added work while balancing the process best we can."

John showed a couple of slides covering the aspect of paper kaizen we were describing and what the overall intent was. "But before we attack our current process, we are going to pick up with round 2 of our airplane exercise from yesterday," John stated. We had the teams break down as they were yesterday and go back to their current state balance charts they created previously. "We need to calculate the takt time for the airplanes. Does anyone recall what the demand was and time we gave you?" I asked. "40 planes in 6 minutes," answered Ace, continuing, "so that ends up being 9 seconds per plane." "Correct," John confirmed.

We instructed the teams to lay the takt time line on their charts and evaluate. You could see the imbalance between operators 1 and 2, but also see that both operators were over the takt time. "So, we have some work to do. You guys highlighted the waste and value-added portions yesterday. The objective of paper kaizen is to remove the waste while balancing out the work to the takt time," added John.

We gave the teams about 20 minutes to do the paper kaizen and try to balance out on a new chart. We then instructed them to use the new balance charts to reconfigure or improve their processes for another live round. "Setup the process based on your new OBC. You can change the process flow, add or take out operators, and change the layout if you want," John noted.

The teams got to work and got creative. They asked questions of us as "customers" if they could change things like a fold or how stringent the "quality check" needed to be. We provided some guidelines and they modified their process. The total work content roughly after removing some waste was around 23-25 seconds, so both team honed in on 3 operators now. Team 1 followed our guidance and balanced their first two lines to takt while leaving the third operator with the slack time. The second team however, led by Ben and Mark, worked the other way and back loaded the cell.

We let them run a round as it was laid out. Both teams failed to meet the output but team 1 got to 33 planes, while team 2 had 28. "We will run one more round so take a few minutes to make some adjustments. You have to hit the demand this round or you lose the business!" John declared.

The third round was a good one for team 1 as they output 41 planes, with only one reject, so 42 in total. The flow was solid and well balanced, with the last operator who had some free time helping with other tasks such material labeling or handling for the other two. Team 2 got 32 planes, but you could note it was a bit more of a struggle. Worse, they reverted to trying to produce small batches, thinking this would be better. You could see the imbalance as operators 2 and 3 waited at various times.

We debriefed and felt good in realizing that most of the group got the picture of paper kaizen, balancing, and flow. Except for a few - Dave had joined us just before this round and while he just observed, had plenty to say and was openly challenging the batching versus continuous flow. Mark jumped on as well, "I don't see much of a benefit. If we work in batches, each individual stays busy. We just need to make sure they produce enough to keep the next person busy as well. That's how we run today and as we saw, most people were busy." "Is busy operators the name of the game?" I asked. "Or is getting product through to our customers as quickly as possible?" I continued. Mark didn't have an answer, just sort of blank stares, and Dave magically had a phone call to take as he left the room, having thrown his grenade. We still had a few folks to convince but the bulk of the team got the picture.

We then turned the teams' attention after a quick break back to the actual XS product charts at the back of the room. "You guys are seasoned now in paper kaizen. We calculated takt time earlier and drew the lines in. Challenge now is to do the paper kaizen work to balance out the XS process best we can," I instructed the team. With that they set out to do the work and balance out the process.

After about 30 minutes the team had done some good work in taking out some waste and reconfiguring the balance charts with value added work. The total work content end to end was down to about 125 seconds, down from the total earlier of 191 seconds. "Estimating the number of operators is straightforward. Take the total work content time and divide by the takt time. 125 seconds into 31 seconds yields 3.8," John said writing on a flip chart. "We will round this up to 4 to start with the ultimate goal of getting this down to 3 long term," he stated.

"This is crazy. We have six people working on day shift and four or five working on night shift today to keep up, with overtime. There is no way we can do this with four people total," shouted Mark. It was time for Ben to chime in. He had been quiet until now. "I agree. This looks goofy. I know we can improve but this seems really aggressive," Ben said in a snarky tone.

"The math doesn't lie. It rarely does," John said. "If we are in line with our process elements and times, we should be fairly close. Close enough to move forward with process design in an innovation round," he added. At that point Greg chimed in, "I've long thought we could improve productivity in assembly. I think this week is confirming that the opportunity exists. I was skeptical when we did the layout 3P and to be honest a little skeptical coming into this that we could achieve what we were targeting. But you can literally see the opportunities and John's right, the math does indicate it's possible. We need to keep going and at least see."

We had had our problems with Greg for sure in the past, but this was Greg at his best. He was well respected in this plant. He had come so far in the last few months of working with him and as a lean practitioner, there is rarely anything more rewarding than seeing the "transition" of an initially resistant plant manager to a lean advocate.

With the group sort of reinvigorated by Greg's words, we moved them to the next stage which would be an innovation round focused on process flow and cell design. The future state or ideal balance charts would continue to be key as we progress and would be modified as we go. John covered a few simple slides on this part of the process. "This is all about creativity. How can we change the process flow and then use that thinking to create an optimal cell design that uses the minimum amount of space, has the best flow, and is the most productive," John said.

I was adamant that we have our engineering friends in this as their input from the product side is key for changing the process. "Where the hell is Dave and Vivek?" I asked Greg. We broke the group into two teams with a nice mix of ops, ME, quality and engineering. But would not start without everyone here. "We aren't starting until they are here and each one on a team. This is too important to not have design side input. Great ideas on the manufacturing side could be bad for the product side. Needs to be joint in the process," I flatly declared. We gave the team a break while we set out to find Dave and Vivek.

I went with Greg this time and so did John. We found Dave and Vivek both in an engineering conference room with two other engineers and we filed in. I asked Vivek and the other two to leave and once they did I said, "Dave, look. We are approaching the half way point of this event and you have spent roughly a total of maybe 90 minutes with us, and Vivek maybe two hours. We need design side support for this to work and you and Vivek were committed to the event. At a minimum, we need Vivek in the event if you are too busy. Chris had worked with Frank to ensure we would have full engineering support for this event and others needed to pave the path to making the layout for this plant work."

Frank was the division VP of engineering and Dave's boss. Dave was a bit surprised that we had approached him and was this direct, but we had had enough. "At the report out this afternoon, Chris and Frank will be on. I asked them to make sure they could be and they both confirmed. I am going to be very clear with them on how much support or lack thereof we have had from engineering. But we are about to start a very important exercise that maybe you guys could join and change that report out narrative," I offered as an out.

Dave was a bit taken back, "Sure, I will be right over with Vivek. We just had some NPD work to do which is why we were busy." I looked at the screen and it was product concepts for a 5-year road map. "This plant won't be here in two years if we don't get all this work done now. I would maybe reshuffle some priorities near term. I know Chris had that conversation with Frank. Looks like you might want to as well." With that we exited and said "See you in 5."

Having hopefully cleared the air, Dave and Vivek both showed up after the break ended. "The innovation round will be timed out at 75 minutes. You guys are basically tasked with evaluating your current state, your ideal balance chart based on that, and then looking to develop some alternative ways to do the process need to end. Once you have a flow concept, use that to design the cell. We are looking to push for as many different options as you can develop," John instructed. "Consider the form, fit, and function of the product as well as quality characteristics we discussed earlier to ensure we preserve that but then anything is on the table. Consider the wastes and opportunities we saw. Everything is on the table," I added.

After the round each team had two potential concepts fleshed out. Like the L3P exercise, they documented both and we posted them up on a wall for a gallery walk. Team 1, led by Greg, had a process flow that went stem sub, valve assembly, engrave, clean, and then test. Team 2 had a flow very similar to today, O-ring prep, valve sub, stem sub, assembly, engrave, clean, test. The teams then had some different layouts that would bring these ops together. Each team walked the group through their flow and layouts. We then set them up with some voting tickets, and using the design criteria of flow, space, and productivity, we had them vote on each. The results were then input into the decision matrix and revealed the best in order.

The top two ended up being the two from team 1, so we had the group focus there. "Let's do a quick plus/delta on both of these to see what we like and dislike about them, and then we can use that information to develop a single one hopefully better," John said. The group went at it and listed out the pluses and minuses. Using that information, they tweaked their designs a bit to detail out a final concept they all felt good about.

"Clean up the design and detail out in terms of dimensions, equipment, - anything else needed in the cell. Based on the new process flow, we may have to reconfigure our balance chart now which is important to keep current," I told the group.

The teams finished up their detail design work on the flow and the cell. Dave and Vivek were present and split on the teams, and as the group came together for the final design, their inputs from the product side were key as was Maria's, the quality manager. Having the operator input was good as well, even though they had some bias on "how it was done in the past" that we had to help them overcome. In the end, the team had a concept they felt good about and we stressed we are only half way there - the mock ups and trials would help us seal the deal. At this point, they were ready for the probably what is the backbone of all this and the most important factor - standard work.

Perspective and Application

Shifting to future state is another pivot point in the DFF process. Understanding accurate demand information is critical and is used to determine the takt time for the system being designed. Comparing this takt to the current state process work done previous tells us where we are at today in meeting (or not meeting) that takt and how our balance looks.

Using the paper kaizen technique to help identify opportunities to take out waste and optimize VA work is a key step to helping to optimize the process as a first step. Teaching the group to look at things in this manner can help them think about process design differently and also help them see how balance can help meet customer demand more smoothly and optimize productivity.

The airplane exercise can be continued for another round to help demonstrate these concepts quickly and simply. It continues the collaborative process in a competitive way, a sort of intra-event ice breaker, and yet helps the teams to continue seeing the application and possibilities.

The innovation round brings it together to help the teams develop ideas for changing and improving the process flow and then using that to develop cell design concepts. Using the timed round approach and decision method from L3P, although shorter and not as rigorous, is a good way to drive some innovative thinking and develop a good final layout option.

The final layout can then be leveraged to "rework" the balance chart in an optimal way and balance out the work to takt accordingly. The teams may not know it but at this point, they are developing the framework for standard work which is coming next.

As you can see in the storyline, having the right support in the group is important. We would not suggest doing some of this core work without the cross functional team as it can lead to challenges later. We have seen innovation rounds go down with no engineering presence and then when they do come back in, they completely rip apart the teams process design and tell them it is not possible due to this or that. That leads to a demoralized team and ends up in destructive rework. Make sure the right representation is in the room even if you have to fight for it.

Chapter 5 Action Summary

1. Determine accurate annual demand numbers and boil this down to a daily demand.

2. Using available time, calculate takt using the formula:

Available Time / Daily Demand

3. Draw a takt time line on the current state balance charts. Look for opportunities and discuss the balance and impacts. Correlate it back to what the team experiences today - batching, waiting, overtime....

4. Have teams conduct paper kaizen and work to best optimize the current balance chart to develop a future state ideal one.

5. Run an innovation round with split teams to develop as many concepts as possible for process flow and cell design.

6. Use the decision matrix to determine best two options and then do a plus/delta review.

7. Have teams use the plus/delta outcome to develop a single concept to move forward with and detail it out. Use the final concept to develop a revised balance chart that is balanced to takt.

Chapter 6: Standard Work - The Most Important Factor

"Time to work on what we describe as the backbone of DFF and what we consider the most important factor, standard work. I'd be hard pressed to tell you that standard work isn't critical in not only sustaining processes but also driving new improvements.," I said, cycling through a couple of slides to the summary page. "How have you guys attempted to drive standard work in the past?" I asked the group.

"It has typically been an engineer or management exercise," replied Jeff flatly. "We really didn't get the operators involved and then expected it to work wonders once we created it," he added. Joe, the assembly manager agreed, as the two operators in the room both nodded. "They would ask some questions and time us, but whatever the engineers would come up with wasn't the way we did things and the time wasn't realistic," Hazel noted. "So, we never really followed it, unless a manager was watching us," Brad said in a quiet tone.

"Exactly, and this what we commonly see all over. Standard work is not an engineering exercise. It should be a team effort to develop what we see as the best way today we know how to do a process, with waste removed and value added work maximized. I saw today, because the point is standard work is living, and should be improved upon over time. That is kaizen and that is how you get improvement and productivity. It's a waste of time if not including the operators, because chances are it won't be followed," I instructed the room.

"It's most critical for DFF as we are designing and implementing a new process, possibly quite different than the old one or something completely new from scratch. If we don't have strong standard work developed, the outcome will not be what we want," John noted. "Fortunately, a lot of the work we have done with the balance charting and design work feeds right into the standard work," he added.

We covered some basic principles and fielded some questions which led to some good discussion. Especially with the Hazel, Brad, and Joe. Mark, our supportive friend, was still not convinced. "This stuff is a bunch of bullshit. We spend all this time writing it, it is not followed, and ends up taking our time since we have to audit and do other stuff to try and make sure it's followed," he said in a loud tone. Greg stepped up, "I think that's what they just said Mark. We haven't been doing this right and need to change our view of it and how we go about it. For this process, we have you, Hazel, Brad, and Joe in here along with other functional leaders. We should be able to work out the "best way" and agree on it," he said.

"Exactly Greg. We will draft it, test it, adjust it, and by the end, be running trials with it to validate it, and can even bring in other operators to help validate it. Once we agree, we sign off on it. Yes, we sign off on it together. In addition, that does not mean it is for life, it can be adjusted after 2 or 3 months if we find improvements. That's the spirit of standard work," I said. "Once operators are using the standard work, the documents become reference for leaders to do spot audits, yes to confirm it is being followed but the intent is to look for opportunities to improve the process. The audit should not be viewed as a policing tool but one to drive improvement," John noted.

With the group better understanding the intent and behavior of standard work, we got into the technical aspects of it. "We calculated takt time previously and this is a key element of standard work, but there are two others," John said, continuing, "One is the sequence of operations which we have spent a lot of time over the last day designing and refining, in your balance charts as well as process flow and cell design. The third element deals with flow, which we have talked about. In order to maintain smooth flow, sometimes the balance to takt is not there between operations, so we need something to cover us."

"Inventory," Ben said in a confident tone. "Correct Ben," John added. "We need what we call standard work in process or SWIP, to provide us a buffer to cover the imbalances in our process so we can maintain smooth flow," John continued. "But this is calculated, and not a random amount of inventory," I interjected to clarify. "Today, as we saw, there are batches of inventory throughout assembly tied to this process. All of that drives waiting and the risk of becoming scrap or obsolete, some of the downsides of producing in large batches," John stated.

"Cycle time is sometimes the last or fourth element. This is the manual work time we have been working with thus far. Knowing this and takt, we can calculate the number of operators, dividing cycle time into the takt time," I said writing the elements on a flip chart with the equation. "We pretty much have all the information we need. We just need to draft the standard work as we have it laid out in our process design, so we can use it to test the process when we do mockups and trials," we added.

With that, we put the group to work in writing out the standard work on the sheets provided. The group was somewhat familiar with standard worksheets, with their past forays into standard work. They pulled elements from the balance chart as well as the times to begin building the standard work. "Do we break up into the number of operators as we write?" Joe asked. "Just write it for now as if one operator was doing all the work. We will have to validate our balancing during trials and verify things so tough to write in hard breaks for operators now. This is a rough draft," I replied. "We also want to look for points that we noted in our work thus far where we have safety or quality concerns. These should be noted at the correct points of the standard work so we build ways to deal with them into our processes when we do mock ups," John said. After about 45 minutes, the team had good working drafts of the standard work available and was ready to move on.

Perspective and Application

Standard work is one of if not the most important factors in your operation. It drives consistency and efficiency that will enhance safety, quality, throughput, productivity, and help regulate WIP inventory. It cannot be over stated how important it is to not only develop it correctly but also to ensure it is being utilized correctly. The problem is most don't really comprehend the true intent and approach it incorrectly. Therefore it typically never sticks.

It's a tool to give us a baseline on the best way we know how to do the process today. It should be regularly reviewed via auditing or by the team using it to find ways to improve it - this is kaizen. For DFF, we view it as the backbone of the whole approach as we are designing new processes or cells, and building (or rebuilding) the standard work as we go from the ground up. It is a collaborative effort by the team and should have operators, team leaders, and supervisors involved.

The core elements: work sequence, takt time, SWIP, cycle time should all have been developed by this point or at least information available via the balance charting or cell design to determine. Developing a rough draft of the standard work now is sufficient as it will be tested, reworked, and validated. A good working draft by the end of the DFF effort is the goal that can then be used for implementation and further refinement.

The auditing may be a policing tool upon initial deployment but over time this should become more focused on finding ways to resolve problems the team is having or ways to improve the standard work. It should not be management versus the operators with standard work but a team effort to find the best way and keep improving it. If you have to police constantly and drive its use from the top down, something is wrong. Maybe rewind your efforts and consider an approach like DFF to help redeploy in a more collaborative and effective way.

Chapter 6 Action Summary

1. Provide training and discussion on standard work focusing on the why, behavioral aspects and then the technical components.
2. Review the elements and discuss information and data available now to develop draft standard work.
3. Write the draft standard work using the balance charting and cell design/process flow worked out before this.

4. This is just the rough working draft. There will be a few evolutions of it through the event and this also shows the team how we go about auditing and improving it in an iterative approach.

Chapter 7: Process 3P - Mock Ups

--And then the Captain said looking of the front of the boat at the wild waters ahead, "now the fun begins." --
Unknown

We were heading into day 3 of the event in Breckenridge. You could say yesterday we were in the bathtub of despair, that dip in the middle of events where people get emotionally drained, sometimes frightened with the challenges encountered. The energy was low as people funneled in slowly, and we had to chase down stragglers like Dave. Vivek was here though. He had been in the event since we had the chat with Dave earlier in the day yesterday. People were looking tired and we had the peanut gallery led by Mark popping questions like would we finish today. "But the hell with that," I said to myself, "today was mock up day and if this didn't energize the team like it had done in past events, then I am not sure what could."

"Today we will be leveraging a 3P technique sometimes referred to as moon shining or mock ups," John stated as we kicked off. "Wait, we are going to repeat what we did in the layout event?" Ben asked in a confused tone. "No, we are going to literally build our work cell today out of whatever materials we have available and bring it it life," John replied. "This is where things get fun and also get real. Up to now we have been doing our current state studies, learning, and then worked up our future state designs. Now we have to build it and validate it as quickly and as effectively as we can," John continued.

We covered some very basic slides on 3P mock ups - mainly to inspire more than anything but also to provide some guidelines on that they would be doing. "This is creativity at its best," I said. "We want to mock up our cell with readily available materials and make it as close to fully operational as we can. As we go, you will be testing your process and standard work, and tweaking things. Once we have a build, then we run trials and continue to refine not only the process and standard work but also the cell itself," I said John cycle through some slides showing mock up examples.

With that we headed out to a clear area in the back of the factory. Jeff and his team had cleared out some space that was a part of maintenance. It provided a clear footprint, about 15 feet wide by 20 feet long, which would be good enough for the team to build up the cell as the footprint by design would be about 9 feet wide by 10 feet long. There were plenty of materials back there as we had worked with Jeff and Ben ahead to ensure there was materials on hand. Card board, some old benches, tools, tape, Creform building materials, Styrofoam, and other associated supplies were on hand. Plenty to get going.

"This will be run timed for two reasons - to drive a sense of urgency and therefore creativity similar to the innovation rounds, and two, to keep us on schedule as we need to be running formal trials by this afternoon and validating the standard work," John commented. "What limitations do we have? I mean, do we have to keep it similar to what we have out there today?" Joe asked while looking over at Mark.

There was an odd dynamic with Joe, the assembly manager, and Mark, the area supervisor. Joe had been "promoted" into the role after his predecessor was let go, and his boss before that had quit. It was a tough job given the environment and Joe was feeling the pressure. Mark had been around for nearly 20 years and was a plant rat here far before Dynet acquired the site. He was one of the old guards, this is the way we have always done it guys. Joe had been around for about 4 years only in comparison. Mark didn't want the manager job. He knew better. Now Joe had it and Mark was not always the most helpful. We could sense Joe's frustration from when we first talked to him about a month ago. You could feel the tension between them this week.

Coming back to Joe's question, I replied, "You have no limitations. You guys have redesigned this process on paper. It's time to bring it to life. Get creative and do it," I stated while looking directly at Mark. Mark whispered something to Hazel who was next to him, and they both started laughing. Mark was the type of person who infected others with his attitude and challenged the culture. We had similar problems with Ben in our previous L3P work here, but Ben eventually came around for the most part. Here, Mark was a front-line leader and clearly wasn't on board. We had hoped through the week he would change course, but he seemed to be just getting worse.

"Something you want to share with the team Mark?" John asked, a bit pissed that Mark was whispering and snickering. "Nothing. I'm just surprised Greg is letting all of us continue to jerk around here all week. Greg, do you know how much our past dues have climbed the last 3 days alone with us in here?" Mark snorted. Greg shook his head, paused, and then answered, "Mark, why don't you excuse yourself from the event and go check on past dues since you have been wanting to find a way out all week. I think the rest of the team can handle this from here. You haven't been engaged at all really so maybe you can do something more productive in assembly."

"I can't believe you are kicking me out of the event Greg?" Mark said in a confused tone. "Yes, I am. And if anyone else wants to join him, now is the time. I have had it up to my eye balls with the negativity, and saying we can't do this, shouldn't do that. I admit, I have been a skeptic in the past with lean but my eyes are wide open now folks and I see the benefits in this. It's not easy but if we do this together, we can make it happen and reap the benefits. We don't have a choice. We have a plan to execute to in source work and if we don't, chances are this site will be closed. So, there's the reality," Greg said in an animated tone.

Mark shuffled out of the area slowly kicking a stack of bins on the way out. He was surprised Greg kicked him out essentially. But Greg pulled his card and had seen and heard enough. I was surprised as well by Greg's response but as encouraged as ever. It was inspiring to see a leader make a stand, make a stand for lean and for doing what's right. "Thanks Greg. I did not expect that but it was overdue. We said at the beginning, this has to be a team effort and we have to stay together to carry through that bathtub of despair we talked about," John commented.

I think Mark exiting the event sort of galvanized the team. I know for damn sure it gained Greg some respect among the group. He sure got some more from John and I. There was a bounce in the step of people as they began their mock ups. We didn't give a ton of guidance on the mock up as it tends to constrain. Just build the cell. And you have the entire morning to do it but you need to be testing and tweaking things as you go. Keep the footprint as small as possible considering the number of operators, equipment needed, and size of the parts as well as SWIP needed.

The team got it. Even Hazel was moving swiftly helping to cut cardboard boxes down with Ellen and Brad. Joe, Ben, and Jeff were working to get the actual tester dismounted from the current large table it was on and remounted on a smaller table to reduce the footprint by half. Greg and Harold were working to mock up material flow racks with bins and focused on part presentation into the cell. Vivek and Ace were looking at the various tools and how we could simplify to reduce the number needed and looking at where they were needed. Dave was even there working, a total surprise, and was with Maria and Mike, a maintenance guy, figuring out how to cut down a six-foot bench to 3 feet. Dave was asking for a sawz-all to do it. Pretty amazing.

The good thing about this was we didn't direct any of this traffic. The team was working to their plan and making it happen. "Sometimes a small ignition can be enough to light a fire, despite the environment" John said as we observed. We observed, offered some feedback on things and fielded some questions. One good one that we had discussed yesterday in the cell design training was a u-shaped cell versus linear. The team had a u-shape design coming out as their top design but was questioning it now as they mocked up. They were also debating going counterclockwise or clockwise flow.

"One of the benefits of a u-shape is reduced walking, it's also a tighter footprint," I noted to the team. "We had talked about that yesterday but linear is not bad if it makes sense from a flow perspective. You can also break down the work and flex the hand off points along the linear line to increase efficiency as far as movement and SWIP," John added. "We tend to flow counterclockwise for a lot of reasons, but most notable is that roughly 90% of the population is right handed as we showed yesterday, so if you step through the motions, it's very likely more efficient flow wise to go counter," I said.

The team continued building up the cell. It was great to see this empty space begin to emerge into a compact looking collection of small tables and card board concoctions. As it came together, Joe, Hazel, and Brad began walking through the steps per the standard work to ensure things were in the right location - parts in order of use, tools where needed, nothing blocking flow, etc. The team made changes on the fly as they went and the mock up was coming together.

By 11 AM the team was beginning to run some slow, methodical trials and continued making refinements. "The flow is looking good thus far," I commented to Greg as we observed. Greg agreed, "Yes, this is really looking good and being able to see it and experiment with it is the nice part about it." The team was broken down into small teams to verify standard work, look for waste or opportunities, and also look at the flow of material and people, just as they did for the current state. They were continuing to note opportunities to continue refining, which they would continue through lunch.

At the tail end of lunch, we gathered the team. I noted, "You guys have been amazing this morning. You mocked up the cell, have run through some early trials and found ways to keep improving what you have. Things look good. This afternoon is about finishing strong. It's time to move to actual build trials to validate and finalize our drafts of standard work, as well as lock down the cell design so we can begin documenting things for the actual implementation to come. Great work thus far. By the end of the day, we want to have our standard work drafted, an updated balance chart, a detailed sketch or drawing of the cell layout as well as benches or racks needed, list of tools and supplies required, and anything infrastructure wise noted on layout we need to plan and execute."

Perspective and Application

The 3P mock up work is one of the differentiators when using the DFF approach. It is also a fun pivot point to go from the future state design work and actually go build things and be able to tinker. It's also here the team can see the design come to life and be able to test it. All teams end up normally rebalancing their work and editing the standard work several times through the day. This is impactful for a couple of reasons, as it helps refine the standard work and optimize it, but also teaches the team on the iterative nature of standard work and how it evolves. In fact, in our experience, going through the DFF process is one of the best ways to demonstrate work balancing and standard work creation.

The mock ups or moon shining can utilize whatever materials are present to build the cell. Sometimes coined "cardboard city", this creative design thinking inspired exercise is intended to drive the creation of a functional cell where you could practically build product. This may not be perfectly possible but the team should strive to get as close to reality as possible. For those steps where you can't fully execute, getting as close to simulation is suggested. Being as "real" as possible for the test runs is strongly advised.

Once built, it's all about testing, and re-testing and adjusting, so on and so forth. Done initially with a single operator, it really becomes tinkering and making small incremental improvements to shave off seconds, reduce additional waste, etc. The team should treat this testing round as they did when doing the current state process study - have time observers, waste watchers, and flow mappers watching actual people build the product. This information is then used to continue refining. Small improvements matter here in the design phase, as it will provide a compounding effect.

All this iteration and experimentation typically leads to a smooth operation and an improved cell that is meeting takt in the most efficient way possible. Updating the balance chart and standard work based on the new process and times observed creates a solid implementation starting point. Once the cell is setup and running somewhat smooth, it may be a good idea to bring in some others not on the team who can observe, offer some feedback, and maybe even run the cell for a few cycles. Getting this outside feedback serves two key purposes, one being to get some broader engagement from others outside the team and help drive support during implementation. Second being, it is always good to get some outside eyes and perspective for fresh thinking and ideas. It has never hurt us when using this approach and has always proved helpful.

Chapter 7 Action Summary

1. Using the cell design and standard work draft, mock up the cell in "cardboard city", bringing the cell as close to reality as possible in scale and ability to actually build product. Use whatever materials are available to complete the mock up.

2. Walk through the standard work in the cell after built and look for opportunities to change or tweak the design such as location of tools or components. Micro improvements matter at this stage.

3. Run trials and experiment. Start slow and have one operator build all the way through, observing the motion and noting any opportunities.

4. Adjust and then begin running formal build cycles. Have team broken down to observe and verify standard work, map flow, and look for waste and opportunities.

5. Adjust and repeat until running full speed. Add in additional operators and work out balancing to meet takt. Continue making adjustments as needed.

Chapter 8: The Details and Plans

The team worked over the next few hours on running additional build trials to validate standard work and times, as well as tinkering with locations of parts, tools, and supplies to optimize. They updated the balance chart as instructed and felt good about what they had around 3 PM. "Spend the next couple of hours documenting everything. Sketch out the cell layout, ensure you get power or air drops, and make sure you note bins, tools, and any other supplies needed," John commented as the team prepared for a break. "You basically want to create a work package to go to do the work to implement the cell as efficiently as possible," I added.

"When will do the implementation?" Greg asked. "You guys need to figure it out here as you build the action plan. We typically advise no more than 3-4 weeks as it gets hard to get dialed back into the work done during the design. I would try to turn as quickly as possible but realize some things may be needed infrastructure wise or some tools may be needed, things like that. That's what the team needs to talk through here now," John advised.

The team divided up to accomplish the crime scene study work to be accomplished. Ben led a few on the infrastructure planning, while Joe led a small team on the benches with Ace helping sketch out, and Harold the materials manager led a few others on getting the flow racks detailed out as well as bins with Jeff helping them. We jokingly refer to it being a crime scene as someone coined it in a past event and it sort of made sense. The team had blistered through a mock up to construct a cell, materials scattered everywhere, and it DID look like the scene of a crime. And now we were asking the team to go through and do a detailed study of it to document it. It did fit the profile of a crime scene investigation!

I snapped back to reality when Ben approached and asked about power drops. "We need a 120VAC drop for the tester, some airlines, and water connection. Where we are putting this in the back NW corner, we don't have freed up yet. There is currently some secondary machining done back there that will be moving next week. I reckon we move those and at same time start making the drops, but two weeks feels more reasonable since my guys are stretched with the other L3P moves," he said. I told him that helps to note as that will likely set the timing for the intermission period as we call it. "No other major equipment or tools with long lead times look to be needed. The internal moves and infrastructure may be the gating item. That will provide some time for the team to prep and get the benches built as well as material flow racks," I advised.

As the end of the day approached, the teams had their documentation completed. The cell layout was finalized with dimensions and drops. The bench and rack dimensions were detailed with material needs, and lists of needed tools, bins, and supplies to buy were completed. "Make sure the standard work and balance chart are clean and current. If you have everything detailed, let's build the action plan with owners and timing," John said.

The team used a simple flip chart to list out key actions and timing. Ben added his tasks he commented on earlier, and other began adding actions, with owners and dates. We challenged some of the timing to build benches. "It cannot take a month to build benches guys," I advised. "We need to move a little swifter and be able to implement sooner. You don't want huge time gaps between the design and implementation as it's tough to get dialed back into the work over time and for the simple fact, you guys probably need to do at least 5 more of these in assembly alone. Ground has to be covered, and your next one of these may be a double," I finished.

Greg agreed we needed to move quicker and the team made some revisions. In the end the infrastructure work including moves was about two weeks to be safe, so the team fell in line and agreed to commit to getting all tasks done in the next two weeks. With owners, the plan was complete and we advised putting this into an action plan template that could be reviewed daily in their team meetings to ensure staying on track. The team agreed to do the implementation event in two weeks, meaning all work, supplies, and anything else needed would be complete and on hand ready to roll.

Perspective and Application

Once everything is in place and verified, the team needs to lock down the design and document it through the "crime scene investigation". As this was a mock up, things need to be documented so it can be implemented as efficiently as possible when the time comes. This can be done through pictures, sketches with dimensions, parts lists, and notes. Between now and the, there may be actions for what we refer to as the intermission period such as preparing space, infrastructure such as power, or buying certain equipment or tools. Having an action plan to get this stuff done and being prepared for implementation will make everything go quicker and easier when the time comes.

The cell details for implementation need to be developed but an action plan also needs to be created leading up to and through the execution of the implementation event. The plan should include anything that is needed to get prepared for implementation - when that week arrives, there should be nothing stopping full implementation and getting ramped up. That week will be about speed and a "flowing" implementation.

The team should also be aggressive in their plans. There should not be a large gap in time between the design and implementation work, as it becomes harder to bring the team back together and get after it. This is especially true of doing multiple events in a plant. In fact, if the team is ready for implementation, in theory, it could be done in sequence after the event, perhaps even the next week. But ensuring the readiness is there and having things needed in place is key. Worst case scenario is trying to do implementation missing key items or tasks required to be done and they aren't. That leads to false starts and hampered implementations that don't fully meet expectations. Preparation wins.

Chapter 8 Action Summary

1. Once cell mock up is verified and meeting takt, lock down the design and begin documenting the work with pictures, sketches, dimensions, and notes to be able to replicate during implementation. Key items to capture:

- Cell dimensions and footprint with infrastructure needs such as power, water, and IT for working into plant layout and location
- Dimensions and materials list for benches, racks, and other structures
- Lists of tools, gages, and equipment needed, especially those that need purchased

- List of bins, correctly sized for components and their locations in the cell

2. Update the standard work draft as well as the balance chart to use as an implementation starting point.

3. Develop a joint action plan with assigned owners and timing to complete. Identify what is absolutely required before starting the implementation ('A' priority) versus needed during or just after implementation ('B' priority). Ideally you want the intermission period to be no more than 3 weeks, ideally 1-2 weeks out over time.

Chapter 9: Wrap Up and Intermission

That evening after wrapping up, we had scheduled a team dinner at a local pub. Nearly the entire team joined, except for Dave and Mark, who had exited the event earlier that day. It was a good dinner and we gave our regards in addition to Greg saying a few words. "It's been a great event thus far and you guys have really shown some dedication in this, learning quick and really applying," John told the group. Greg echoed, "I am really happy to see the progress. It's even more encouraging to think about the possibilities when you envision this across all of our assembly and test area, perhaps even the entire factory. Really proud of each one of you for joining this and looking forward to us now going and applying it more."

Dinner was a good unload period for the team as people chatted over drinks. It was part celebration, but also important team building as we talked in a side bar with Greg, Ben, Joe, and Jeff. "This is important. You guys should find the time to celebrate the small wins like this, but also get people interacting outside of the tasks at hand. This team building may seem small, but it is absolutely critical as you push through periods of sometimes tough change," I shared with them. Greg agreed and acknowledged it had been a while since they had done anything like this, probably the last was the smaller team dinner we had during the L3P event.

"We still have work to do in the morning before we wrap up. We must leave around 11 for flights, so want to tie up loose ends and ensure you guys are good to go. The implementation event is in three weeks," John noted. "We will setup a weekly call to ensure things are staying on track, but you guys have to be meeting and reviewing daily and course correcting," I added. "At times, coming out of these, the team gets off track with the intermission work and have run into not being ready for implementation. That never goes smoothly," I continued. Greg committed to ensuring he would be on top of it daily and confirmed, "we will be ready."

The next morning, we worked with the team to clean everything up and prepare a report out. There would be a pitch at 10 AM, inviting some other team members from the site but also allowing Chris and Frank to call in. A few slides were put together and we helped the team create the summary of potential improvements:

- Overall footprint for XS production from 987 square feet to less than 212 square feet - **improvement of 79%**
- Throughput time improvement for assembly, test, and packing from 6 days to less than 4 hours (single order) - **improvement of 91%**
- Productivity improvement - # of total operators from 6 (12 over two shifts) to 4 total (2 per shift) - **improvement of**

67%. Elimination of OT (40 hours per month on average) and needed only for flexing to spikes in demand.

- Quality improvements to improve First Time Through from 81% to >95%... **Improvement of 14%** and helps speed throughput

"Some really solid breakthrough improvements," Greg commented at the slide. "This is our highest volume product line, its growing but the margins are slim. This will be absolutely loved by the sales team. Us getting quality product through faster at a lower cost," he continued. "It's definitely a good story to tell. Execution of it is needed but the design looks great," John confirmed.

The report out went well. Some of the Breckenridge team we brought in asked some good questions, which for the most part the team handled. Frank, the engineering VP, asked about design team input and Dave was quick to chime in that he and Vivek were in the event and provided the needed support. They feel good about the process changes. He also brought up the new XS5 valve line that is in development, which the team did discuss yesterday and how that can actually start off production in this cell and as it ramps, a duplicate cell can be implemented to handle that volume. Frank and Chris both loved that.

After the pitch, John and I packed up and headed out. We shook hands with the team and said we would talk next week and now the time to execute is key. We thanked them for their hard work and said we looked forward to seeing this put in place.

Over the next few weeks, we had our periodic calls with the team each Thursday, and each progressive call, the team was working the action items and Greg was making sure of it, focusing on the A priority items getting done first. The team was executing the intermission plan with the discipline and rigor needed, and on track for the implementation event.

Perspective and Application

Tying up loose ends is important, nearly as important as bringing the team together to celebrate the hard work and bond a bit as in reality, the work has just begun. Summarizing the effort and being able to tell the story to others is key to getting support, not only from the other workers impacted but also leadership. If done properly, the results projected should be breakthrough in nature, meaning double digit improvements in key measures.

This is another transition point for the team. We mentioned the bath tub of despair previously, and this should be the light at the end of the tunnel we referred to. The team should have begun ramping upward as the design came together in the mock ups and trials, seeing the detail plans, and now seeing the summary pull together and helpfully the energy experienced during the report out. This should end the event on a high note.

Chapter 9 Action Summary

1. Prepare a detailed summary of the design, plans, and improvements. Tell the story through a good report out, sharing the work but also accepting feedback to keep improving.

2. Have a team celebration, such as a dinner to help bring the team together and celebrate the hard work completed as well as prepare them for the work to come.

3. Ensure a cadence is established during the intermission period - reviewing progress on site daily and then some weekly pulse to make sure there is traction and readiness.

Chapter 10: Bring It Home - Implementation

We had agreed with the team on two weeks for the intermission period to get the prep work done, so the implementation event could go ridiculously smooth. But we had noticed on the third week's call some hesitation and waffling with some of the remaining actions. "Are the benches and racks built?" John asked, knowing this was a key activity leading into the event. "We are working on them and will try to get them done before Tuesday," Ben said in an unsure voice. The team in the room did some bantering in the background and we could hear, "I thought you were supposed to do that", "No, you were, see it says it on the action plan..."

"Greg, have you guys been reviewing things daily as we discussed?" I asked him. "We have but slipped a bit this week. We had some production issues that caused some challenges so the team got a little side tracked. We will get focused today and get what needs to be done completed," he answered. "So we have some work to do with the structures. Will we have the tools and fixtures on hand and ready by Tuesday, Wednesday at the latest?" I asked. After a pause by the group, they replied with a few voices saying "yes".

That Thursday call did not go like the previous two and John and I both knew the situation. We had been through this before. Heading into an event with the team not fully on track with the intermission hit list. This could go south and we could end up struggling to build and fabricate structures, move equipment, drop power lines, scramble to get tooling on hand, or things could come together at the start and it could go smooth. If the earlier, we should have just done the implementation on the heels of the design event as that is typically the scenario. If the latter, then should be good.

We ended the call in an air of uncertainty and said we would be there Tuesday morning at 7 AM to meet and go through things. The implementation event is all in the shop - no training (done in design event), and all the materials including standard work and balance charts is brought to the work area. It truly is about implementing fast to get the cell built up and running as quickly as possible to minimize down time.

On Tuesday, we showed up as planned and found Greg and Ben immediately. We exchanged some banter and then headed to the work area. Greg and Ben began preparing for what was to come. "So, we didn't get a chance to get all of the structures built. One of the fixtures is still being made in the tool room and we will try to have it done tomorrow," Ben said. "I have my guys here dropping the power lines today," he continued in a sullen voice.

"Guys, we stressed during the design event and over the last few weeks how important it was to have some of these key actions done and ready to go. That's the whole point of doing this in phases so when we come back for the implementation, we can fly through it and be producing almost to rate buy the end of the week. You guys aren't ready. Why?" I asked, trying to maintain my composure.

By this time Joe had shown up to hear my inquiry as well. And our disappointment grew. The three of them began blaming others, blaming each other, and in general, not taking any accountability for the situation. "Stop. Please stop," John interrupted. "You guys are just creating excuses now. We fortunately put buffer build time into day 1 of the implementation events for this reason. But I don't know what is more disturbing, the fact that you guys don't have the actions done or are trying to make excuses and blame others as to why. Greg, you are the site leader and have to own it all. Let's get the team together in the room and let's take an honest look at where things are and break down the group into teams to attack it." Greg felt the need to reply, "Your right. I sort of let things get away here and should be accountable. I'll be sure to remind the team on this that we missed and need to be better."

We rallied the team in the conference room, pulled the action item list up, focusing on A items. "No bullshit here. We need an honest assessment on where things are for each of these. Some items aren't ready and we need to play catch up today," John stated bluntly. Greg walked through each item and got a rundown form each owner. Of the remaining items, a few benches and racks needed to be built, one power line needed dropped, and the fixture needed to be completed. We divided the teams to focus on each area and said we would meet in two-hour intervals for a checkpoint.

It took the better part of the morning and a few check points, but by lunch time we felt we had a handle on the situation and were "caught up". It could have been worse and the situation could have drug into Wednesday. John refocused the team as they finished lunch, "We are in a better place now than we were about 5 hours ago. But we still have some work to do here to get the cell fully stood up and running. Our goal is to have this cell operational but the end of the day, even if that means a long day."

The team got it, and Greg, where accountability had been fleeting in the last week leading up to today, stood up again and began rallying the team. With the team re-energized they set out to finish up the tasks at hand, with Greg leading the charge. The team did get after it that afternoon and although it was a late day, they got the cell stood up and operational.

The next morning was debug and trial. We had some team members get the MDI system and hour by hour board in place, while another team did a walk through per the standard work to ensure everything was in place Joe asked about 5S and visual management, things like shadow boards. "Don't lock anything down right now. You guys will likely be tweaking things over the next few weeks. Just use temporary things like painters' tape and wet erase marker to put things in place. Can almost guarantee a lot of things will shift around a bit," I answered.

We had Hazel do the single operator walk through and we adjusted things as she moved through the cell. Once we were confident things were good, we began running half speed trials. "No timing here. Just focus on building the product per the standard work with one operator. We are looking for issues to clean up before we go full speed," John advised. We did find some things to adjust during the half speed trials, and after all the choreography, after lunch we were ready to to run full speed trials.

We started with single operator just to validate the builds and check timing against single operator balance. Being close after a few cycles, we moved to two operators and had Brad join in. It was cumbersome at first, but we worked it out. "The hand off point was a bit odd at first." Joe advised and then made some suggestions to improve it. Once tweaked, the two-operator balance was validated and we ran some product, being within roughly 10% of the target cycle time.

"Let's add a third operator, not required per demand now, but may be in the future," John asked. We put Joe in as the third builder on the back end. The third operator would not have much to do but did alleviate the finishing tasks as well as packaging. It ramped the output up considerably and based on this, we felt the cell was capable of handling any projected growth over the next year at least, with a single shift.

Thursday was the same, more tweaking and more flat out running production with two operators, Hazel and Brad who as they learned and practiced picked up pace and were hitting takt nearly consistently. Yes, there were still issues but they were being captured real time on the hour by hour board and the team actively worked to resolved them real time using a root cause counter measure approach. The cell was fully operational Thursday afternoon and hitting target output. Although parts were quarantined initially, quality was confirming them for use through the validation process.

The team prepared a summary report out for Friday morning and we did their gemba walk with the leadership staff to visit the MDI and review the hour by hour, advising how to run the cadence and practice the lean oriented leadership needed to properly support. For the report out we brought the entire assembly area to the cell and showed it to them running while Greg and team did the pitch. It was a great gemba oriented report out and the team drew some great applause for the hard work. The report out call was held simultaneously, and Chris and Frank were brought in via the call and a webcam, to try and experience it. Both were happy and amazed at the transformation.

As we packed up and prepared to head out for the airport in the conference room, Greg approached. "Guys, just a heads up we let Mark the assembly supervisor go today. We had been having some issues with him since the L3P, and his behavior in the DFF event, combined with his attitude after that just sealed the deal. We have decided to promote Brad into the supervisor role," he said. "That's a great move. Brad is a proactive, humble leader with great potential. He will do well," I said. John added, "Yes, it's tough with Mark. Some folks just have trouble letting go of old ways and opening their mind up. Lean isn't hard but it also isn't easy. But it is about mindset. If you can't change your thinking, you won't get it. Hope the best for him."

"When will you guys be back?" Asked Ben as he walked up. "Not sure. You guys have quite a bit of work to do and you got the tools now to do it. We are a phone call away if you ever need any support," John answered. With that, we thanked each team member one more time and gave our regards before heading out. It was a great way to end the week indeed.

Perspective and Application

The implementation part of these efforts should go smoothly if the design work is complete, accurate, and the action plan with at least key items is completed headed into the event. One of the benefits of splitting the design and implementation efforts is to allow the team time and energy to focus on creating an optimal design, and then using that to build an implementation path. The implementation event should be all working in the gemba, laser focused on getting the final cell or line built up and operational as quickly as possible so trials can run, bugs worked out, training completed, and the cell operating at near peak output by the end of the week.

But sometimes the prep work isn't done as was seen in the case with XS cell in Breckenridge. This happens and if it does, the team must scramble to get things situated on day one to recover and keep the implementation on track. This is why a good intermission plan is key and driving a daily pulse on the activities. It sort of defeats the purpose of this to not have the key tasks done during the intermission - we mind as well just kept going after the design event and fought through the implementation work.

If things do go to plan or the situation is overcome, the outcome is typically the cell or line that the team worked hard to design, hitting the projected targets or close to it, sometimes even exceeding them. This is incredibly encouraging for the team and the business. Achieving these types of breakthrough improvements is always positive, builds confidence, and the teams likely want to plan the next one soon, as was the case with Dynet.

Chapter 10 Action Summary

Implement, implement, test, improve, re-test, improve, train, and run.

1. Finish up any "open" key actions from the intermission plan. For this reason, we advise leaving a buffer on day 1 of implementation week to allow time to finish building structures or any remaining infrastructure work needed.

2. Implement the cell or line. Most often, you will be replacing an existing line or series of production of areas, which must be taken off line and "repacked" as sometimes equipment is to be re-purposed. This must be navigated to cover production needs - if buffer stock can't be built, the line may have to be switched over on night shift. Implementation also includes putting tools, supplies, and other items in place with basic 5S principles and visual management.

3. Test and re-test. Run trials as was done in cardboard city to verify the standard work and update it. Continue refining.

4. Train new operators. Have event team members train the operators on the standard work and the cell. Then run trials with them getting up to speed.

5. Implement MDI system and hour by hour tracking board. Also finalize any safety, quality, or management of change audits needing completed prior to release.

6. Run. Have team broken down to observe from outside the cell and have the cell just run for a few hours as if normal production using the hour by hour to track outputs and issues. Have team act as team leader to help resolve issues to keep production going.

7. Finalize improvements - last day of event, cell should be running production ideally with minimal interruptions as the event comes to an end. The cell should be at least 80% to takt attainment, ideally closer to 90%. Over time, as the team gets more comfortable and the cell "works itself" in, the numbers should improve and meet or exceed takt output.

Chapter 11: Afterword

The list of lean or process improvement tools is endless. Likewise, the knowledge pool on lean and how to use it to improve systems and processes is endless as well. Sometimes having a structured approach to bring things together in a cohesive fashion is beneficial to drive a focused effort.

That is the intent behind Design for Flow. It's not creating new tools or methods. In fact, many of the tools individually have been around for some time. But what it does do is create a guide lined approach to using the right tool at the right time in right manner. In doing so, teams can achieve some dramatic breakthrough improvements rather than the more traditional incremental improvements when evaluating existing processes and making some minor changes. Incremental improvement is great and necessary, but sometimes a bit more radical thinking is needed to take things to another level. This could be when performance is lacking and significant gains are needed, or for a new product where a new and innovative process is required.

DFF fits the bill there and can also help with the approach, can also help with breaking down cultural challenges and getting teams to work closer together and be more creative. As we saw in the Breckenridge story, there will always be some in the group who oppose thinking outside the box and using "different" tools to drive different thinking and improvements. Sometimes a structured approach designed to foster team work and innovation helps, but at the end of the day, leadership, as we saw with Greg, is critical to building a lean oriented culture that embraces change. And sometimes, tough changes as we saw with Mark, are needed to keep moving forward.

We hope you find the DFF approach or a variant of it useful for your operations. The impacts and potential are real and can help improve safety, quality, productivity, and space utilization dramatically. When combined with Layout 3P for macro level factory flow and space utilization, the DFF approach can really help dial in micro level flow and connectivity in a team based, collaborative fashion in the pursuit of the ultimate objective of continuous flow.

Chapter 12: Appendix - Resources

Typical DFF Event Flow

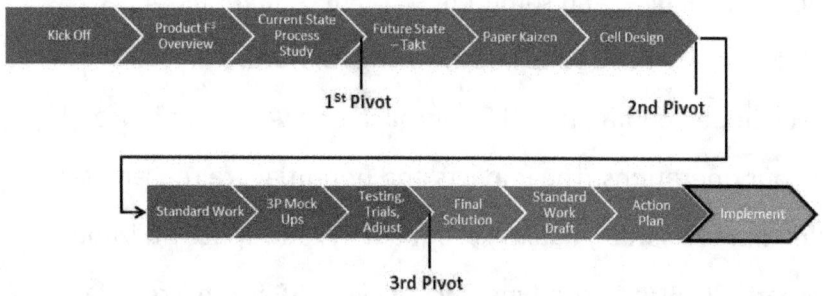

Figure 5: Typical DFF Process Flow

General Information or Terms

L3P, Layout 3P – Production Process Preparation (3P) technique focused on macro flow through facility layout as well as optimizing space utilization.

8 Wastes – The non-value added work in manufacturing or business processes. They include the following:

1. Defects – Scrap or rework
2. Over Production – Producing more than customers buy
3. Waiting – Time spent waiting for information, parts, etc
4. Non-Utilized Talent or Ideas – Self Explanatory
5. Transportation – Excess movement of materials
6. Inventory – Raw, WIP, or Finished Goods

7. Motion – Excess movement of people

8. Over Processing – Doing more work than customers require

7 Flows – Conceptual view of various flows in manufacturing to consider. Allows us to appraise the current state and help prime thinking for future.

1. Flow of people (motion)

2. Flow raw material (transportation)

3. Flow of WIP (transportation)

4. Flow of finished goods (transportation)

5. Flow of information

6. Flow of equipment (material handling equipment as example)

7. Flow of engineering or tools

7 Ways – Innovation driver, challenges the team to come up with 7 (or more) different alternative options during the design rounds. Sometimes the team achieves it or more, but the point is to strive to get the maximum amount of concepts. It is typically used in earlier rounds during divergent thinking.

PQPR – Part Quantity Part Routing, a matrix (shown below) to help analyze and prioritize focus in terms of grouping products into value streams as well as determining parts to align into work cells.

Needle Valves		\multicolumn{7}{c}{Product Quantity / Process Route (PQPR) Analysis}	Prepared	Date
\multicolumn{2}{c}{Product Family}								John D	25-Mar
Part Number	Part Name	Demand Quantity	% of Total	\multicolumn{7}{c}{Process Routing}	
				Machine	Turn	Deburring	Sub-Assy	Marking	Final Assy	Test	Package
NV-2-1	1/8" Needle Valve Std	1000	43%	1	2	3	4	5	6	7	8
NV-4-1	1/4" Needle Valve Std	650	28%	1		2	3	4	5	6	7
NV-6-1	1/2" Needle Valve Std	450	19%	1		3	4	5	6	7	8
NV-2-2	1/8" Needle Valve Spec	125	5%	1	2	2	3	4	5	6	7
NV-4-2	1/4" Needle Valve Spec	75	3%	1		3	4	5	6	7	8
NV-6-2	1/2" Needle Valve Spec	35	1%	1		2	3	4	5	6	7

Figure 3: Example PQPR Matrix

VSM – Value Stream Map, a tool used to map out physical product and information flow end to end. Example shown below.

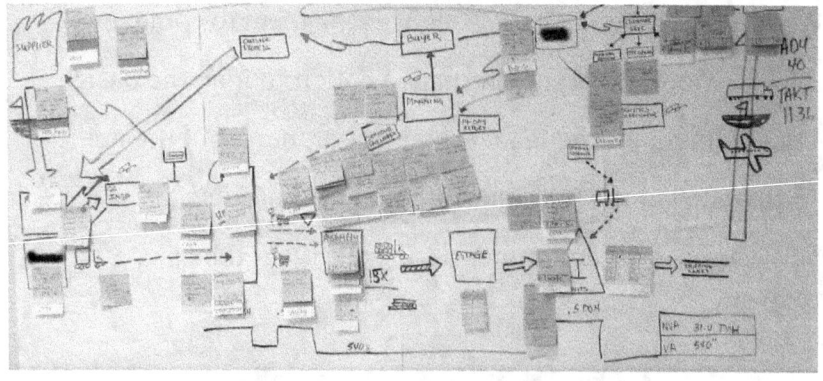

Figure 4: Example Value Stream Map

DFMEA – Design Failure Modes and Effects Analysis, a tool to analyze a product design to identify failure modes in terms of impact and probability of detection and occurrence. Identifies critical design characteristics to plan for in manufacturing, application, and service.

PFMEA – Process Failure Modes and Effects Analysis, a tool to analyze the manufacturing process of a product to identify failure modes in terms of impact and probability of detection and occurrence during production. Plans for critical characteristics and high-risk process steps to develop robust processes.

Work Element – Smallest increment of a unit work in a task. In theory, an element can be transferred from one operator to another.

Time Observation – Observing and documenting work elements and then timing them to obtain real task times. Used to obtain cycle times in processes.

Takt Time – The rate at which customers are consuming product. Calculated as available time divided by demand in units.

Balance Chart – Visual chart that shows the work balance. Time is along the Y-axis and process steps or operators are along the X axis. Takt time line is drawn on the chart to show balance to takt.

Paper Kaizen – Method to analyze process balance charts and work element breakdown to remove waste, optimize sequences, and rebalance work to create a better process.

Cell Design – Process of designing the flow and cell layout using innovation round format with 7 ways logic.

Standard Work – Documented best way of executing a process today. Consists of takt time, work sequence, and in process inventory.

3P Mock Ups or Moonshining – Method to mock up the cell design out of readily available materials and bring it to life in order to test and conduct trials.

Page left intentionally blank

www.ingramcontent.com/pod-product-compliance
Lightning Source LLC
Chambersburg PA
CBHW071218220526
45468CB00002B/661